ISBN 978-1-332-13377-2
PIBN 10289262

**English**
**Français**
**Deutsche**
**Italiano**
**Español**
**Português**

# www.forgottenbooks.com

**Mythology** Photography **Fiction**
Fishing Christianity **Art** Cooking
Essays Buddhism Freemasonry
Medicine **Biology** Music **Ancient**
**Egypt** Evolution Carpentry Physics
Dance Geology **Mathematics** Fitness
Shakespeare **Folklore** Yoga Marketing
**Confidence** Immortality Biographies
Poetry **Psychology** Witchcraft
Electronics Chemistry History **Law**
Accounting **Philosophy** Anthropology
Alchemy Drama Quantum Mechanics
Atheism Sexual Health **Ancient History**
**Entrepreneurship** Languages Sport
Paleontology Needlework Islam
**Metaphysics** Investment Archaeology
Parenting Statistics Criminology
**Motivational**

# GUIDE

### FOR THE

# PENNSYLVANIA RAILROAD,

### WITH AN

# EXTENSIVE MAP;

### INCLUDING THE

# ENTIRE ROUTE,

### WITH ALL ITS

## WINDINGS, OBJECTS OF INTEREST, AND INFORMATION USEFUL TO THE TRAVELLER.

PHILADELPHIA:
T. K. AND P. G. COLLINS, PRINTERS.
1855.

# DIRECTORS

## PENNSYLVANIA RAILROAD FOR 1855-56.

---

### By the Stockholders.

J. Edgar Thomson,  
C. E. Spangler,  
George W. Carpenter,  
Washington Butcher,  

John Yarrow,  
Wm. R. Thompson,  
John Farnum,  
William Neal.  

### By the City of Philadelphia.

George Howell.

### By the Commissioners of Alleghany County.

William Robinson, Jr.,      Thomas Scott.

### By the Board.

William B. Foster, Jr.

| | |
|---|---|
| President . . . . . . . . . | J. Edgar Thomson. |
| Vice-President . . . . . . . | Wm. B. Foster, Jr. |
| Treasurer . . . . . . . . | Thomas T. Firth. |
| Secretary . . . . . . . . | Edmund Smith. |
| General Superintendent . . . . | H. J. Lombaert. |
| Chief Engineer . . . . . . | Herman Haupt. |
| Solicitor in Philadelphia . . . | Job R. Tyson. |
| Solicitor in Pittsburg . . . . . | William A. Stokes. |

# GUIDE

FOR THE

# PENNSYLVANIA RAILROAD.

---

THE city of Philadelphia may be regarded as the eastern terminus of the Pennsylvania Railroad, though the road itself terminates at Harrisburg. The importance of the road is at once felt, when it is understood that it is the only direct highway between the cities of Philadelphia and Pittsburg.

A brief statement of the extent and resources of Philadelphia will suffice to show that she is second to no city in the country in business importance.

In the first place, the city of Philadelphia contains, as shown by the last census, a greater number of dwellings than any other city in the Union, Exceeding New York by twenty-three thousand six hundred and one dwellings. No inconsiderable town in itself.

In the next place, it is the metropolis of a commonwealth increasing in wealth and population at a greater ratio than any of the old thirteen States. In the last apportionment for the House of Representatives at Washington, the State of Pennsylvania gained a member of Congress and the State of New York lost one.

Philadelphia sends out her railroads and canals in every direction. On the north and west she reaches within easy distance her *own coal fields;* deposits which are inexhaustible, and which contain the only anthracite coal yet discovered in America, and from which all American markets are supplied. Philadelphia is the chief seat of this trade; a circumstance to which she owes the inestimable advantage of cheap fuel, which has been an efficient cause of her vast manufacturing superiority. The wharves of the Reading Railroad at Philadelphia, one of the principal places for the shipment of coal, are a curiosity in themselves, well worthy the attention of a stranger.

In manufactures, Philadelphia is far before any other city in America. Every variety of article, large or small, civil or military, intended for use or ornament, is produced here. She is the centre of trade for the manufacture of gas-fittings, her large establishments in this department being the wonder of all who see them. Nowhere else are wrought-iron tubes for conveying gas made on the same scale. The manufacturers of leaden pipe for the whole country reside here. In the production of paper-hangings and umbrellas it has long been conceded that no other city approaches her. One manufactory for paper-hangings extends an entire square in length. On the corner of Fifth and Cherry Streets, stands an immense and symmetrical structure, a monument to the taste, enterprise, and great business energy of its proprietors, in which several hundred hands are constantly employed in the production of the most beautiful fabrics, military goods, coach trimmings, laces, ribbons, embroideries, &c. &c. Every imaginable article composed of iron is manufactured in this city. The locomotives of Philadelphia are known on every railroad in the United States, and on many in Europe; and the car-wheels produced here are not less esteemed. It is said of one of our locomotive establishments, that it has the *largest* and *best* arranged workshops in this country, and when fully occupied employs over 1400 hands, and it has turned out *three* complete locomotives *in one week*. Its capacity is fully equal to 150 per year. The steam-engine manufactories of this city for stationary engines, and for steamships, are known over the world. The engines for the steam-frigate Mississippi, among the few really successful ones in the service of the United States, was made at a well-known establishment in Philadelphia. The manufacture of superior stoves is also carried on to a very large extent.

In the lesser articles made from iron, such as axes, saws, pitchforks, cutlery, surgical instruments, nails, screws, hinges, she is without a rival.

Paper of all kinds, woollen goods, prints, calicoes, and especially carpets, and boots and shoes, are staple productions; and it is affirmed that there are more fine shoes and boots made in Philadelphia annually than in all New England. Carriage-building and harness-making are also carried on very extensively; in both these departments, Philadelphians carried off the prize at the London exhibition.

The Delaware bounds Philadelphia on the east, while the Schuylkill winds its way through the heart of the city. By the former, Philadelphia is put in communication with the ocean; by both, in addition to her railroads, she finds access to the rich mineral and agricultural products of the interior of Pennsylvania, Maryland, New York, and of the far West. The coasting trade of this city is the largest of any of the American cities. Of the amount of her foreign commerce no true notion can be formed from the business transactions at *her own* custom-house. The facilities of transportation from New York are such as to induce Philadelphia merchants to avail themselves of the proximity to the ocean of that port for the introduction of their goods; immense quantities of which are transported at once, with the original case unopened, except so far as the revenue officers require them to be opened, from the ship in New York to the counting-house of their owners in Philadelphia.

Philadelphia has for years been considered the great distributing mart of the United States, and it is not too much to say that the north side of Market Street exhibits some of the most magnificent stores that have ever been erected.

Philadelphia is the medical capital of the country. She has not less than five first class medical colleges of the old school; while the homœopathic, the water cure party, and the vegetarians have all flourishing schools. The medical literature of the country centres here. While her law, miscellaneous, and school-book publishing houses are among the largest in the United States.

If the traveller is not bent on business, but is in pursuit of elegant recreation, let him stop in Philadelphia, and he will be amply rewarded. The Academy of Natural Sciences, on Broad Street, next to the La Pierre House, is an institution unapproached by any other in America. The second, if not the first, ornithological collection in the world is there. It contains the late Dr. Morton's invaluable collection of skulls, certainly the most extensive collection in the world, besides a valuable scientific library, and numerous rare specimens in zoology and mineralogy. The Wagner Institute, lately incorporated, contains wagon loads of collections for the study of mineralogy, geology, and botany. On Fifth near Chestnut Street stands the Philadelphia Library, founded by Dr. Franklin, one of the oldest and most extensive libraries in the country. It contains nearly 65,000 volumes. Independence Hall, in Chestnut above Fifth Street, where the Declaration of Independence was signed, adorned with the original and remarkably exact portraits, taken during their lives, of most of the distinguished characters, American or foreign, who took part in the events of our revolutionary period, offers attractions to the patriot and the student of history, such as no other place can present.

The Pennsylvania Hospital, it might almost be called world-renowned, and the Eastern Penitentiary, also known throughout the best portion of the world, draw hither many visitors. The Fairmount Water Works and Laurel Hill, and the beautiful summer excursion on the Schuylkill, between the two, have enchanted many strangers.

On Chestnut Street below Broad Street, stands the beautiful building appropriated to the United States Mint, where the curious and nice operation of coining may almost always be seen. And further down Chestnut Street, above Tenth, the Academy of Fine Arts offers to the visitor at all times a delightful entertainment. Many of the largest productions of the well-known Benjamin West are there, and there only, to be seen; among them the famous picture of Death on the Pale Horse.

The Girard College, principally composed of marble, is the grandest building in America, and the most richly *endowed* charitable institution, by a single individual, in the world. The Custom-house, the Farmers and Mechanics' Bank, the Bank of North America, the Girard Bank, and the Exchange are public buildings which would attract notice in any part of the world.

The eccentric John Randolph, of Roanoke, when in Europe, wrote to his friends in America that nothing impressed him so much as the grandeur of the European architecture. "It seems," said he, "as though I had never seen anything in America but huts—unless it were the Capitol at Washington and the Bank of Pennsylvania, in Philadelphia."

The new gas works of the city, on the Schuylkill, below Gray's Ferry, are on a scale never before attempted in this country, if in any other. Those curious in such matters could not do better than pay them a visit.

The neighborhood of Philadelphia produces in abundance a delicate and luxuriant grass, of which cattle are very fond, to which is ascribed the superiority of the milk, butter, and cream used here—a superiority so marked that the most unobservant traveller never fails to notice it. In like manner the beef of the Philadelphia market is always preferred to any other. The hotels are not, as a general thing, on the gigantic scale of the New York houses. They are, however, quite as comfortable, and always offer to the stranger a table spread with delicacies from a market proverbial for its excellence.

But the steam-whistle gives token that the iron-horse is ready for a start, and we must hurry away; but, before we part, we can assure the traveller that no matter how often he repeats his visit to Philadelphia, we can continually exhibit to him new scenes and matters of interest. Her charitable institutions, for instance, of which we have made no mention in the foregoing sketch, except the Girard College, embrace the greatest variety of objects, are endowed with exceeding liberality, and are conducted with a degree of intelligent faithfulness and zeal which make them models. Her public schools also are of the first order.

The accompanying map contains the red line of the road from Philadelphia to Pittsburg, with the towns and stations, streams and mountains, accurately located. At Harrisburg, it will be readily perceived that, in order to prevent the map from being too wide, the Susquehanna River was cut in two, where it was pursuing too much of a northerly course, and the western section commenced at the bottom of the map. The windings of the road, in order to climb the Alleghany Mountains, may be distinctly seen.

The line of road from Philadelphia to Pittsburg has three owners. First, The state of Pennsylvania owns that part extending from the city to Dillersville, one mile above Lancaster, consisting of a double track, in length 69 miles. At Dillersville, the Harrisburg, Portsmouth, Mount Joy and

Lancaster Railroad commences, and continues from thence to Harrisburg, a distance of 36 miles. At Harrisburg, the Pennsylvania Railroad commences, and completes the line from thence to Pittsburg, a distance of 248 miles.

The cars are drawn from the depot by horse or mule power, out Market Street, and across the Schuylkill Permanent Bridge, at the west end of which they take the locomotive. The river Schuylkill was, until lately, the western boundary of the city, as laid out by its founder, William Penn. The act of consolidation of 1854 extends the boundary westwardly to Cobb's Creek, a distance of 3¼ miles. The waters of this beautiful river rise in the mountains of Schuylkill County, which contain the treasures of anthracite coal, peculiar to this State, millions of tons of which annually find their way to tide water on its bosom, and upon the Reading Railroad along its banks, whence it is carried to other cities and towns, in several States of the Union.

After taking steam we pass up the Schuylkill in full view of the light and graceful WIRE BRIDGE on the right, the FAIRMOUNT WATER WORKS, and the beautiful fall of water over the DAM, as well as the placid sheet which it makes as far as the eye can reach. The new bridge at Girard Avenue may also be seen, and the GIRARD COLLEGE, with snowy whiteness and its magnificent marble columns and marble roof, overlooking the city and surrounding country for miles. The State locomotive engine house is immediately on the road to the right, a few hundred yards from the place of starting. Thence passing through a deep cut we curve round and pursue nearly a westerly course, leaving the city and its busy multitudes behind. In rounding the curve to the left we may observe the West Philadelphia Water Works (now belonging to the city), being a very high iron column cylinder, encircled by a neat and tasty iron stairway, winding around it from its base to its summit. At a distance of 3 miles from the depot we pass HESTONVILLE on our left, then LIBERTYVILLE, and ATHENSVILLE, and arrive at WHITE HALL, 10 miles from the city. Just before arriving at this station, we may observe, to the left, a large building with an extensive lawn, and a handsome wood between it and the railroad. This is the Haverford College, belonging to an association of Friends, and conducted by them, where a classical education may be obtained by the youth of that denomination, but which is not confined exclusively to them. This college is in Delaware County. It was at Chester, on the Delaware River, in this county, where William Penn landed in November, 1682, with his cargo of English Quakers, in the ship Welcome. There were Pemberton, Moore, Yardley, Waln, Lloyd, Pusey, Chapman, Wood, Hollingsworth, Sharpless, Rhoades, Hall, Gibbons, Bonsall, Sellers, and Claypoole, whose ancestor, Cromwell, not many years before, ruled the destinies of the British empire. The descendants of those men have become numerous in Philadelphia, Chester, and Delaware counties. The birthplace of the celebrated Benjamin West is a few miles south of the Haverford school—his ancestors accompanied Penn on his second visit to Pennsylvania, and were also Quakers. Benjamin was reared in the faith and profession of his ancestors—a profession from which he never swerved when his genius commanded the flattery of courts, and honor from kings and princes. It is recorded of him, by Galt, that at the age of seven he made a drawing, in red and black ink, of an infant niece, of whose cradle he had the charge, and whose sweet smile in her sleep excited his imitative powers, though he had never seen a picture or engraving. With this precocious sign of inherent talent the boy's mother was charmed, and her admiration and encouragement confirmed his taste. At school, even before he had learned to write, pen and ink became his cherished favorites; and birds, flowers, and animals adorned his juvenile portfolio. It is a tradition of the family that the father, having sent Benjamin out to plough, missed him from his work, and found him under a cokeberry bush, where he had sketched the portraits of a whole family so strikingly that they were instantly recognized. Another anecdote is related of him—that one day at Rome (where he had gone to complete his studies), while his master had stepped out a moment, West slyly painted a fly on the work on which his master was engaged. The master came in, resumed his work, and made several attempts to scare away the fly. At last he exclaimed: "Ah! it is that American." We next pass the stations of VILLA NOVA (a Roman Catholic college), MORGAN'S CORNER, and the EAGLE, and arrive at the PAOLI, 20 miles on our journey. The train frequently stops here for refreshments. Near this place 150 Americans, under General Wayne, were killed and wounded on the night of the 20th of September, 1777, by a detachment of English under General Gray. This action is frequently called the Paoli massacre. General Wayne was surprised in the night by a superior force, and no quarter shown. "On the 20th of September, 1817, being the 40th anniversary of the massacre, a monument was erected over the remains of those gallant men by the Republican Artillerists of Chester County, aided by the contributions of their fellow-citizens. It is composed of white marble, and is a pedestal surmounted by a pyramid. Upon the four sides of the body of the pedestal are appropriate inscriptions." The country through which we have passed is thickly dotted with neat farm-houses and barns, and all sorts of comfortable out-houses for pigs, and poultry, sheep, cattle, and horses. The large fields of grain and grass which greet our eyes in the summer season, the herds of cattle, and flocks of sheep, everywhere to be seen, indicate great agricultural thrift in the inhabitants of Delaware, Montgomery, and Chester counties, through which we have been passing. The small whitewashed stone houses which we may observe at a short distance from the dwellings, and generally situated under the outspreading branches of some ancient oak or willow, with a crystal brook stealing away through the luxuriant grass, are spring-houses. We may observe the patient cows standing around, with their white udders swollen with milk, waiting to yield it to the milk-maid's pail, from which it is poured into earthen or tin pans, and those are placed in the clear cool water of those houses where the rich cream is formed for the butter. From these houses is taken the far-famed Philadelphia butter, superior to that, it is said, of any city in the world. The secret of its superiority lies in the green grass peculiar to this rolling country, and the cool springs of water that rise from its hills. No prairie land, how rich soever it may be, can ever produce butter equal to that made in the rolling counties around the city of Philadelphia. Soon after passing the Paoli, we commence descending the valley hill, and now breaks upon the view of the passengers one of the most picturesque scenes that can be found on any road in the Union. This is the celebrated Chester County limestone valley. This valley extends easterly and westerly some 20 miles in length, and averages 2 miles in width. It is skirted on both sides with high hills covered with timber, from which issue innumerable springs of pure water, converted into perpetual fountains in the valley, and affording a never-failing supply for man and beast, at the house and barn. This valley is noted for its fertility and beautiful farms. As the cars descend the hill, on an easy grade, the passenger may take in at one view many miles of this magnificent panorama, interspersed with comfortable and neat farm-houses, spacious barns, and other necessary buildings. Hundreds of fields of waving grain, the deep green corn, and luxuriant timothy and clover, pass in review before him. Here, the farmers may be seen driving their "teams a-field," and there, cattle, horses, and sheep, feeding in the pasture, or reclining under the trees. This valley supplies the finest beef for the Philadelphia and New York markets. The cattle are brought when poor from the regions of the north and west, and fattened here in the rich pastures of Pennsylvania. The beef of Philadelphia, like the butter, is nowhere else to be found. After running along the southern side of the valley for several miles, and passing the STEAMBOAT, OAKLAND, and the intersection of the Valley Railroad leading to Norristown, we arrive at DOWNINGTOWN, a distance of 33 miles from the city. This is a quiet little rural village, to the right, which has grown up on the Philadelphia and Lancaster turnpike, which passes through it. The north branch of the Brandywine Creek, over which the cars have just passed, also flows through it. This stream commemorates the battle of Brandywine, which took place in 1777, between the English, under General Cornwallis, and the Americans, under General Washington, at Chad's Ford, some 15 miles below this place. Its pure waters, that flowed between the contending armies, were dyed by the blood of the combatants, and bore witness to their deadly struggles. And if the palm of victory be awarded to the side that made the most widows and orphans on that bloody day, as is usually the case, we must admit that the English won the prize. The fiercest part of the battle took place around the Birmingham Friends' meeting-house, a few miles farther up the stream. The combat raged furiously at this place, and many lives were lost. Missiles of war, and other evidences of the battle, were long after to be found here. The thunder of the cannon, and the roar of the small-arms *without* were in strange contrast, as a mode of settling disputes, with the still small voice *within*, by which the Friends profess to be guided in adjusting differences with one another. But however justifiable we may think the cause of it to be, still we must confess that

> "War is a game, which,
> Were their subjects wise, kings would not play at."

Soon after leaving DOWNINGTOWN, and passing GALLAGHERVILLE (a small village to the right), and the CALN station, the road crosses the valley to the northern side, and passes over the west branch of the Brandywine, near Coatesville (a small village to the left) on a bridge 75 feet high, and stretching across a chasm 850 feet. The character of the country is still the same. Agriculture is here carried to a high state of perfection, and every "rood of ground maintains its man." We come now to MIDWAY, a small neat village and station near the high bridge. Pursuing our course on the north side of the valley, we pass CHANDLER'S and stop at PARKESBURG to water, 44 miles from the city. This is a neat village of 400 inhabitants, containing a large hotel and the machine shops of the State. It is the offspring of the Columbia Railroad, and is the residence of its superintendent. Chester County, through which we are now passing, is one of the most fertile in the State. The bounties of Providence are here distributed with an unsparing hand. In 1850, there were raised in this county 1,339,446 bushels of corn, 547,498 of wheat, 1,145,712 of oats, 96,315 tons of hay, 2,092,019 lbs. of butter. The quantity of corn surpassed that of any county in the State except Lancaster. This county is the birthplace of Chief Justice McKean, who was also Governor of the State for nine years. It was also the birthplace and residence of General Anthony Wayne. The train continues on the north side of the valley, which is here sensibly diminished in width, and passes PENNINGTONVILLE and CHRISTIANA, two thriving villages which have sprung up within a few years on the railroad. CHRISTIANA, which is just over the line in Lancaster County, has been rendered conspicuous by a riot that occurred there in 1851, occasioned by some Maryland slave owners attempting to arrest their runaway negroes. The latter having tasted the sweets of liberty for some time, and being joined by their friends in the neighborhood, resisted their claims, and resolved to fight for their liberty. The melancholy result was, that a Mr. Gorsuch, from Maryland, was killed on the spot, and one or two others were severely wounded. A number of those who desired the freedom of the negroes, of both white and colored, were indicted for treason by the United States, but the trial resulted in their acquittal. We now come to the GAP, so called on account of the opening in the mine hill at this place, through which the railroad passes. This is the highest point above tide water on the State road, being 560 feet. At Christiana we diverge from the beautiful Chester County valley which we have traversed a distance of 20 miles, and at the Gap we enter the no less beautiful PEQUEA VALLEY of Lancaster County. Passing KINSERS, LEMON PLACE, GORDONSVILLE, and BIRD-IN-HAND, through an unsurpassed agricultural section, and crossing the beautiful CONESTOGA, we enter LANCASTER CITY, 68 miles from Philadelphia, 38 from Harrisburg, and 286 from Pittsburg. Probably no country in the world can present a finer picture of agricultural prosperity than that through which we have passed from Philadelphia to Lancaster. Taking 20 miles on each side of the railroad between those cities, we have 2720 square miles. There being 640 acres in a square mile, the quantity in the whole would be 1,740,800 acres. This would probably average $85 per acre, making the value of such a strip of land nearly ONE HUNDRED AND FORTY MILLIONS OF DOLLARS. The railroad between those cities cost about FOUR MILLIONS OF DOLLARS. Before this road was built, this *strip* might have been purchased for FIFTY MILLIONS. LANCASTER CITY is the fourth city of Pennsylvania with respect to population. It contains about 15,000 inhabitants. It is the capital of Lancaster County, which is not equalled in the value of agricultural productions by any in the State. The oldest turnpike in the United States has its western terminus at this place, and connects it with the city of Philadelphia. It was incorporated as a city in 1818. The inhabitants of this city, owing to the facilities afforded by the railroad, have of late years become aroused from their lethargy, and have commenced the manufacture of various articles. They have one or two anthracite furnaces in full blast, several large cotton steam factories, locomotive works, and various other manufacturing establishments. They have also water works to supply the citizens with a constant supply of fresh water, and have become so far refined as to have their streets and their houses lighted with gas. A new county prison has been erected at a cost of $110,000, and a new court-house at a cost of $100,000. The inhabitants are chiefly of German descent. In 1850, Lancaster County, which contains only 950 square miles, produced 1,803,312 bushels of Indian corn, 1,365,111 of wheat, 1,578,321 of oats, 96,134 tons of hay, and 1,907,843 lbs. of butter. The crop of oats excelled that of any other county in the United States; that of wheat, any other except Monroe County, New York; that of corn, any other county in the State. This county is famous for its large barns and large fat horses. At DILLERSVILLE, 2 miles west of Lancaster, we leave the State Railroad (which continues on to Columbia) and take the Harrisburg and Lancaster road, which intersects the former at this place.

LANDISVILLE, the next station, is merely a stopping place. MOUNT JOY is a small village in a rich agricultural district. Near this town is a flourishing Female Seminary. Elizabethtown station is near the old village of that name on the turnpike, to the right of the railroad. At CONEWAGO, we cross the creek of that name, which forms the boundary of Lancaster and Dauphin Counties. MIDDLETOWN, the next station, is a place of some note. It is situated at the mouth of Swatara Creek, on the Susquehanna, 9 miles from Harrisburg. There the Union Canal intersects the Pennsylvania Canal. The town contains 1,200 inhabitants, and has a bank and a newspaper office.. Leaving Middletown, we continue our course on the banks of the Susquehanna, which now presents its broad surface to our view, interspersed with islands and eddies, and rafts of lumber from the pine regions above. The canal from Columbia to Pittsburg will now form a conspicuous object on a great part of the route. The sound of the boatman's horn may frequently be heard echoing down the valleys and over the hills, reminding us of the lines of the poet—

O! boatman, wind that horn again,
For never did the listening air,
Upon its joyous bosom, bear
So wild, so soft, so sweet a strain!
What though thy notes are sad and few,
By every simple boatman blown,
Yet is each pulse to nature true,
And melody in every tone.

We now arrive at HARRISBURG, the capital of Pennsylvania, beautifully situated on the east bank of the Susquehanna River, 105 miles from Philadelphia, and 248 from Pittsburg. The Capitol is a handsome brick building, occupying a conspicuous position on a lofty eminence, which affords a delightful view of the broad Susquehanna, studded with verdant islands, and spanned by handsome bridges, with the Kittatinny Mountain in the background. Its population is about 9,000. The valley in which this town is situated, is fertile and productive, yielding its hundredfold reward to the husbandman. Harrisburg has applied to become a city. Being the capital of the large State of Pennsylvania, it has long enjoyed an intercourse with our most intelligent citizens from all parts of the State, rendering its society quite intellectual. Several railroads centre here, giving facilities for communication with all parts of the State. Besides the railway to Pittsburg and Philadelphia, there is the Dauphin and Susquehanna, the York and Cumberland, and the Cumberland Valley, which are finished. Then there is the North Central, up the Susquehanna to Sunbury, and the Lebanon Valley Railroad to Reading, not yet completed. The inhabitants of this capital have also caught the spirit of the age, and are in a fair way to render their town conspicuous for manufacturing purposes. They have also their water-works and gas-works. At this place commences the Pennsylvania Railroad, which we now take, and pass rapidly through a highly cultivated section, having the river on one side and the canal on the other, to ROCKVILLE, where we cross the Susquehanna on a beautiful bridge, 3,670 feet in length, 51½ miles above Harrisburg. The large building on the high eminence to the right, about 1 mile from the town, is the STATE LUNATIC ASYLUM, recently erected, reflecting much credit on the State for thus humanely providing for this unfortunate class of our population. The passengers, in crossing this great bridge, should not fail to take a view of the scenery from about its centre. It is of the most magnificent character.

The Susquehanna Bridge was commenced in 1847. The first contract was abandoned, and the masonry re-let to Messrs. Helman & Simons, of Harrisburg, through whose energy, aided by the exertions of Robert McAllister, Esq., of Juniata County, the whole was completed by December, 1848. Some of the stone was boated by McAllister for a distance of 50 miles, and delivered ready dressed for the work. The superstructure is upon the Heur plan, and was erected by Daniel Stone, who also built the new bridge across the Schuylkill at Market Street. In March, 1849, six spans of the bridge, while in an unfinished condition, were destroyed by a tornado, which was so violent as to blow the water from the river to a height of 30 feet, and carried away timbers 40 feet long and of great weight, by the force of the wind alone.

On the western bank of the river may be seen the Blue Mountain, which, commencing in the southern part of the State, near the Maryland line, continues in an irregular course to the northeast, broken by various streams which have forced a passage through it. The celebrated Delaware Water Gap is found in this range, the whole of which is characterized by its bold and rugged escarpments, and the hardness of its rocks.

At the west end of the bridge, the traveller enters the county of Perry, the surface of which is broken by numerous ridges and valleys, having a general parallelism to the Alleghany Back Bone range, but of limited extent and of irregular character. The lands in this county are not so productive as those which are found east and south of it. The prevailing geological formation of the valleys is the red shale, while the mountains exhibit the usual characteristics of the sand stone, which are found upon every ridge of any considerable elevation in the State.

The Cove station is so named from a very peculiar curve in the range of the Cove Mountain, the summit of which recedes from the river opposite the town of Dauphin, and approaches it again at Duncannon.

The COVE station, 10 miles from Harrisburg, is not a point of any business importance, but it furnishes an abundance of pure water, and wood from the mountain sides.

The next object of interest is the point of rocks at the northern extremity of the Cove Mountain; here the road is forced close to the edge of the river, and a deep cut is encountered, which has given much trouble, from slips of rock, but which is now safe. The county road winds along the mountain at an elevation of sixty feet above the railroad; it was cut out of the solid rock by the Pennsylvania Railroad Company, who were required to make a new road in place of the old one, occupied by the bed of the railroad.

On the opposite side of the river in Dauphin County, is seen the bold elevation called Peters' Mountain, once apparently connected with Cove Mountain, but now separated by the Susquehanna.

Passing the point of rocks by a sharp curve to the west, the traveller soon crosses the beautiful SHERMAN'S CREEK, and his ears are saluted with the heavy reverberations of the forge, the roar of the water-fall, and the busy noises of the rolling mills. The evidences of industry and thrift are conspicuous in this locality, and the enterprise of WILLIAM LOGAN FISHER has erected here a monument durable as brass. Pig, bar, rolled, and hammered iron, nails and spikes are the products of these works. Anthracite coal is brought by canal from the Shamokin region, and various localities, some of which are very distant, are laid under contribution for the best varieties of ores.

One mile above Duncannon is seen the mouth of the **Juniata**; its precise locality may be distinguished by the smaller of the two bridges, which are upon the right. The large bridge crosses the Susquehanna, and is used as a common road bridge, and also as a towing path, by means of which the boats navigating the main line and the Susquehanna division of the Pennsylvania Canal, are enabled to cross the river in the pool of water known to lumbermen as *Green's Dam*.

Duncan's Island, for many years previous to the decease of Mrs. Duncan, a well-known and much frequented place of summer resort, is bounded by the Juniata, the Susquehanna, and a small channel on the north side connecting these two streams. It embraces several hundred acres of excellent farming land, and a considerable population is found upon the island brought together chiefly by the lumber trade, which is very active at this place during the freshets of the Susquehanna. Dealers in lumber from the principal cities and from the various towns on the river below Harrisburg, resort to the island to make their purchases, and secure the pick of the market.

Duncan's Island was a pleasant spot, and those who in former years made it their resort for recreation, and who have enjoyed the society of the hospitable mistress of the mansion, and of the amiable Julia, who have sat beneath the foliage of the majestic forest trees, or rambled along the streams or in the gardens—

> " When opening roses breathing sweets diffuse,
> And soft carnations shower their balmy dews :
> Where lilies smile in robes of virgin white,
> *The thin undress* of superficial light,
> And varied tulips show so dashing gay,
> Blushing in bright diversities of day,"

will be saddened by the recollection that these pleasures are past never to return.

Passing along a rocky side hill for a distance of nearly two miles, the train usually stops at the AQUEDUCT. Here passengers for the Susquehanna region are transferred to the packet-boat, which, after receiving its cargo of human freight, crosses the Aqueduct, and is towed by horses to Williamsport, on the west branch of the Susquehanna.

Leaving our friends in the packet to the enjoyment of their bilge-water and mosquitoes, and to all the comforts of narrow berths, crying babies, and the chances of suffocation, enjoyments which a shower of rain is sure greatly to enhance, we will bid adieu to the blue hills of the Susquehanna, and its broad shining waters, and wend our way to the sources of the gently flowing Juniata, where they gush forth in copious streams from the broad bosom of the Alleghanies.

The next station is BAILEYSBURG. Mahoney Ridge and Limestone Ridge approach the river near this point from the west.

NEWPORT is a thriving town, six miles northeast of Bloomfield, the county seat; it is a place of considerable business, the shipments being large both by railroad and canal. It is the second town in population in Perry County. The Little Buffalo Creek empties into the Juniata at Newport.

Half a mile above Newport is the terminus of Middle Ridge, and at the distance of a mile the railroad crosses the Big Buffalo Creek by means of a *stone* viaduct of five spans.

MILLERSTOWN is the last station in Perry County; it is readily distinguished by the neat station-house, erected by the Pennsylvania Railroad Company, at the end of the bridge.

Perry County is bounded on the northwest by the Great Tuscarora Mountain, through which the Juniata forces its way about one mile above Millerstown, leaving the deep rocky channel worn

> "By ever flowing streams—arteries of earth,
> That, widely-branching, circulate its blood,
> Whose ever throbbing pulses are the tides."

At the Tuscarora Mountain the railroad, the general course of which for several miles has been but a few degrees west of north, turns suddenly to the southwest, following for several miles the base of the mountain.

Eight miles north of Millerstown was formerly a remarkable natural phenomenon, known as the roaring spring, the waters of which rose with great *force below* the surface of the river, causing a violent ebullition accompanied with noise. This spring is no longer visible; the unsparing progress of improvement has entombed it beneath the rocky embankments of the road bed.

We are now on the north side of the Tuscarora Mountain, in the valley of the same name, and in the County of Juniata.

MEXICO, forty-five miles from Harrisburg, is not the celebrated city of the Aztecs, and has never been distinguished by any of the sanguinary battles which marked the progress of Scott's campaign; but it is not without its historical associations. Opposite Mexico, some white men, in attempting to dig a cellar for a house, were shot by Indians; and a terrible war was once waged between two tribes, which originated in a quarrel between some Indian children about grasshoppers, and was known as the Grasshopper War. Sensible, was it not? Quite as sensible as the origin of many of the wars which have distracted nations, boasting of greater advances in civilization and refinement.

PERRYSVILLE, the depot of Tuscarora Valley, is at the mouth of Tuscarora Creek. A flourishing Academy is located in the valley, about eight miles distant.

One mile above Perrysville, the road passes along the face of Law's Ridge, an extremely steep and rocky bluff, consisting of variegated shales and sandstones, which present a most singular appearance, the strata being bent and curved in every direction, sometimes forming semicircles.

From the point of Law's Ridge, the town of MIFFLIN, the county seat of Juniata, is in view, on the opposite side of the river, and in a few minutes the cars stop at PATTERSON HOUSE, built by Messrs. Fallon & Wright, of Philadelphia, and is now owned by the Pennsylvania Railroad Company. They intend to make a first rate house here to accommodate travellers. The town of PATTERSON does not improve so rapidly as was anticipated, although one of the machine shops of the Pennsylvania Railroad Company is located at this place; very few improvements have been made except by the Company. The row of brown frame houses on the hill south of the road, was built by the Company for the accommodation of its workmen.

Four miles west of Mifflin, the road enters the gloomy and contracted passage between the Black Log on the south, and the Shade Mountain on the north, known as the Long Narrows. Previous to the construction of the road, there was no habitation on the south side of the Narrows, and but one public house on the north side. This house was in such a dilapidated condition that, it is said, the engineers, who made the location, were compelled in rainy weather to sit up in bed with umbrellas hoisted, to protect themselves from shower baths, which they had not stipulated for as a part of the accommodations to be furnished. We are not able to inform our readers whether an extra charge was made for the unusual comforts, but it is supposed that "mine host" furnished the baths gratuitously.

At the west end of the Narrows is the large and flourishing town of LEWISTOWN (61 miles from Harrisburg, 187 from Pittsburg), the most important in the valley of the Juniata, the county seat of Mifflin County, and the outlet of the great and fertile valley of Kishicoquillas. Iron ore of excellent quality abounds in this county; there are many iron works, and in the limestone districts several curious caves containing stalactites. The lovely valley in which Lewistown is situated, is a specimen of many similar valleys that adorn this mountain region. It is not to be compared with the valley of Wyoming, of which the poet has sung in the following strain :

"On Susquehanna's side, fair Wyoming !
Although the wild flower on thy ruined wall,
And roofless homes, a sad remembrance bring
Of what thy gentle people did befall ;
Yet thou wert once the loveliest land of all
That see the Atlantic wave their morn restore.
Sweet land ! may I thy lost delights recall,
And paint thy GERTRUDE in her bowers of yore,
Whose beauty was the love of Pennsylvania's shore !"

The valley of Kishicoquillas has not had the advantage of being celebrated in song, like its fair sister on the Susquehanna. But although no Campbell has, in poetic page, rendered this valley forever classical, and thrown around it the witcheries of its own equally lovely GERTRUDES, it has had, nevertheless, what one who knew him well has described, the best specimen of humanity he ever met with, *either white or red*, viz : LOGAN, the celebrated Indian chief. This noble Indian is identified with the valley of Kishicoquillas in which Lewistown is situated, and, although an unlettered red man of the forest, his celebrated speech will be perpetuated in history as long as the beautiful poem, the GERTRUDE OF WYOMING, shall live in classic lore. " A resident of the place boasts, not without some reason, that many circumstances concur to make Lewistown a desirable resort for strangers. The scenery is the finest in the world ; we breathe the pure mountain air. Our clear streams abound with fish, particularly trout. Our forests are filled with game of every description; and Milliken's Spring, on a farm adjoining the town, is ascertained to possess all the medicinal qualities of the Bedford water, particularly in bilious complaints."

Of Kishicokelas,* the Indian tradition has preserved little but the name. Another friendly chief distinguished in American annals, had his cabin for a number of years beside a beautiful limestone spring, on Kishicoquillas Creek, a mile or two above the wild gorge, where the creek passes Jack's Mountain. This was Logan, the Mingo chief, whose eloquent speech is familiar to every one. Logan was the son of Shikellimus, a chief of the Cayugas. Mingo, or Mengwe, was the name given by the Delawares to the Iroquois, or Six Nations.

The following anecdote was related by Judge Brown, the first actual settler in Kishicoquillas Valley, and for many years one of the Associate Judges of Mifflin County :—

* Often spelled in this way.

"The first time I ever saw that spring," said the old gentleman, "my brother, James Reed, and myself, had wandered out of the valley in search of land, and finding it very good, we were looking about for springs. About a mile from this we started a bear, and separated to get a shot at him. I was travelling along, looking about on the rising ground for the bear, when I came suddenly upon the spring; and being dry, and more rejoiced to find so fine a spring than to have killed a dozen bears, I set my rifle against a bush, and rushed down the bank and laid down to drink. Upon putting my head down, I saw reflected in the water on the opposite side, the shadow of a tall Indian. I sprang to my rifle, when the Indian gave a yell, whether for peace or war I was not just then sufficiently master of my faculties to determine; but upon my seizing my rifle and facing him, he knocked up the pan of his gun, threw out the priming, and extended his open palm towards me in token of friendship. After putting down our guns, we again met at the spring and shook hands. This was Logan—the best specimen of humanity I ever met with, either *white* or *red*. He could speak a little English, and told me there was another white hunter a little way down the stream, and offered to guide me to his camp. There I first met Samuel Maclay. We remained together in the valley a week, looking for springs and selecting lands, and laid the foundation of a friendship which has never had the slightest interruption.

"Mr. Maclay and I visited Logan at his camp, at Logan Spring, and Mr. M. and he shot at a mark for a dollar a shot. Logan lost four or five rounds, and acknowledged himself beaten. When we were about to leave him, he went into his hut and brought out as many deerskins as he had lost dollars, and handed them to Mr. M., who refused to take them, alleging that we had been his guests, and had not come to rob him; that the shooting had been only a trial of skill, and the bet merely nominal. Logan drew himself up with great dignity, and said: ' Me bet to make you shoot your best—me gentleman—and me take your dollar if me beat.' So he was obliged to take the skins or affront our friend, whose nice sense of honor would not allow him to receive even a flask of powder in return.

"The next year," said the old gentleman, "I brought my wife up and camped under a big walnut tree, on the bank of Tea Creek, until I had built a cabin near where the mill now stands, and have lived in the valley ever since. Poor Logan," and the big tears coursed each other down his cheeks, "soon after went into the Alleghany, and I never saw him again."

The following additional incidents highly characteristic of the benevolent chief were related by Mrs. Norris, a daughter of Judge Brown.

Logan supported his family by killing deer, dressing the skins and selling them to the whites. He had sold quite a parcel to one De Young, who lived in Ferguson's Valley, below the gap. Tailors in those days dealt extensively in buckskin breeches. Logan received his pay, according to stipulation, in wheat. The wheat on being taken to the mill was found so worthless that the miller refused to grind it. Logan was much chagrined and attempted in vain to obtain redress from the tailor. He then took the matter before his friend Brown, then a magistrate, and on the judge's questioning him as to the character of the wheat, Logan sought in vain to find words to express the precise nature of the article, with which the wheat was adulterated, but said that it resembled in appearance the wheat itself. "It must have been *cheat*," said the judge. "Yoh," said Logan, "that very good name for him." A decision was awarded in Logan's favor, and a writ given to Logán to hand to the constable, which he was told would bring him the money for his skins. But the untutored Indian, too uncivilized to be dishonest, could not comprehend by what magic this little piece of paper would force the tailor against his will to pay for the skins. The judge took down his own commission with the arms of the king upon it, and explained to him the first principles and operations of civil law. "Law good," said Logan; "make rogues pay." But how much more simple and efficient was the law which the Great Spirit had impressed upon his heart, *to do as he would be done by.*

When a sister of Mrs. Norris (afterwards Mrs. Gen. Potter) was just beginning to learn to walk, her mother happened to express a regret that she could not get a pair of shoes to give more firmness to her little step. Logan soon after asked Mrs. Brown to let the little girl go and spend the day at his cabin. The cautious heart of a mother was alarmed at such a proposition, but she knew the delicacy of an Indian's feelings, and she knew Logan too, and with secret reluctance, but apparent cheerfulness she complied with his request. The hours of the day wore very slowly away; night approached, and the little one had not returned. But just as the sun was going down, the trusty chief was seen coming down the path with his charge, and in a moment more the little one trotted into her mother's arms, proudly exhibiting a beautiful pair of moccasins on her little feet, the product of Logan's skill.

Such was the man whose whole family was afterwards barbarously murdered on the Ohio below Wheeling, by some white savages, without a shadow of provocation. It was not long after that act, that his consent was asked by a messenger with wampum, to a treaty with Lord Dunmore, on the Scioto in 1774. Logan delivered to the messenger the following speech which is now well authenticated to have been his own, and not composed, as had been suspected, by Mr. Jefferson.

"I appeal to any white man to say, if ever he entered Logan's cabin hungry, and he gave him not meat; if ever he came cold and naked and he clothed him not. During the course of the last long and bloody war, Logan remained idle in his cabin, an advocate for peace. Such was my love for the whites, that my countrymen pointed as they passed and said, 'Logan is the friend of white men.' I had even thought to have lived with you but for the injuries of one man. Col. Cresop, the last spring, in cold blood and unprovoked, murdered all the relations of Logan, not even sparing my women and children. There runs not a drop of my blood in the veins of any living creature. This called on me for revenge. I have sought it. I have killed many. I have fully glutted my vengeance for my country. I rejoice at the beams of peace; but do not harbor a thought that mine is the joy of fear. Logan never felt fear. He will not turn on his heel to save his life. Who is there, then, to mourn for Logan? Not one."

Logan was the son of Shikellimus, a Cayuga chief, who dwelt at Shamokin in 1742, and was converted to Christianity under the preaching of the Moravian missionaries. Shikellimus had a high esteem for James Logan, the secretary of the province, and most probably had his son baptized with the Christian rites by the missionaries.

McVeytown, also called Waynesburg, is situated on the canal 11 miles above Lewistown. It has been incorporated as a borough.

At Newton Hamilton, the river makes a horseshoe bend, while the road cuts across the neck greatly reducing the distance, but encountering considerable rock excavation. At the west side of the bend the road crosses the Juniata on a bridge 70 feet above the water, and at a considerable elevation above the canal and aqueduct. The river at this place forms the boundary between the counties of Mifflin and Huntingdon.

Mt. Union, 85 miles from Harrisburg and 162 from Pittsburg, in Huntingdon County, is at the entrance of the gap of Jack's Mountain. Passengers are conveyed in stages to Shirleysburg and other towns in the southern part of the county.

The gap of Jack's Mountain presents a peculiarly wild and rugged appearance. The sides of the mountain are almost entirely destitute of vegetation, and covered with immense masses of gray and time-worn rocks. At the foot of the mountain is the village of Jackstown. A few years since, it was announced in the newspapers that an awful conflagration had entirely destroyed the village of Jackstown, not *a single house* having been left standing. The calamity would not have appeared quite so distressing if the writer had stated the fact, that the only house in the village was *a double house*, built of stone, the roof of which was consumed.

Mapleton is the depot for Hare's Valley, on the north of which is the range of the famous Sideling Hill, well known to travellers on the National Road between Hancock and Cumberland.

West of Sideling Hill is the Broad Top Mountain, an isolated elevation containing a small and singular bituminous coal basin, the veins of which are from one to four feet in thickness, of good quality for steam generating purposes. A railroad is in progress of construction, connecting this basin with the canal and railroad at Huntingdon.

Mill Creek, ninety-one miles from Harrisburg, is the depot of the agricultural products of the west end of Kishicoquillas Valley.

South of Mill Creek is a most singular topographical formation, called Trough Creek Valley, formed by Sideling Hill and Terrace Mountain, which unite at the south side of the river in a ridge of sufficient elevation to turn the courses of the streams to the south; after many miles, the waters are again returned to the north by uniting with the Raystown Branch, which empties into the main Juniata a short distance above Mill Creek.

Although the scenery along the Juniata is characterized by great beauty, it is probable that no portion of it is more diversified and picturesque, than that which is presented to the eye of the traveller in the vicinity of Huntingdon. Retracing his journey in imagination, he will

> " Hail in each crag a friend's familiar face,
> And clasp the mountain in his mind's embrace."

Huntingdon County is surpassed by few, if any, in the variety and richness of its mineral deposits, or in the fertility of its valleys. This county, previous to the separation of Blair, contained sixteen furnaces, twenty-four forges, and one rolling-mill.

Huntingdon County was first settled in 1749; its early history is similar to that of the neighboring counties, being characterized by Indian depredations, and feats of endurance and valor on the part of the settlers.

The stream which empties into the Juniata at the town of Huntingdon, is called Standing Stone, from a remarkable stone, for many years held sacred by the Indians, and regarded by them as a sort of talisman.

HUNTINGDON, ninety-seven miles west of Harrisburg, is the county seat. It was laid out previous to the Revolutionary war, and named after Selina, the Countess of Huntingdon, who had been a liberal patroness of the University of Pennsylvania.

Large quantities of lead ore were obtained in this county previous to the war of the Revolution, chiefly from Sinking Valley, on the south side of the Little Juniata, but the low price at which lead can be brought from the western States, where it exists in much greater abundance, has caused the abandonment of the mines in this locality.

At PETERSBURG, the next station, the canal and railway, which have been close companions for more than one hundred miles, part company. The canal takes the course of the Frankstown Branch, through Alexandria, Water Street, and Williamsburg to Hollidaysburg, while the railroad follows the rugged path cut for it by the Little Juniata.

> Here mountain on mountain exultingly throws,
> Through storm, mist, and snow, its black crags to the skies;
> In their shadows, the sweets of the valleys repose,
> While streams, gay with verdure and sunshine, steal by.

TYRONE CITY is the next station on the Little Juniata, at the mouth of Little Bald Eagle Creek, and one mile west of Tyrone forges (owned by Lyon, Shorb & Co.). This flourishing town has sprung up since the construction of the Pennsylvania Railroad, and is rapidly increasing. It now contains about one thousand inhabitants, several first rate hotels, a foundry, a machine shop, planing mill, churches, &c. A plank road, five miles in length, leading to Bald Eagle furnace, commences here. And charters have been obtained for railroads to connect New York and Lake Erie with the Pennsylvania Railroad at this point. It is already one of the best stations on the road—commanding a large share of the business of Clearfield and Clarion Counties.

At Tyrone City the railroad leaves the deep gorge of the Little Juniata, which it has followed through the mountains for the last twelve miles, and enters Tuckahoe Valley. This valley lies between the main range of the Alleghanies and Brush Mountain. Its general width is about three miles, and its length about fifteen miles. The southern side is rich limestone land, and the northern side is clayey soil; the railroad runs nearly through its centre. On the south side of this valley are extensive deposits of iron ore, from which are drawn the supplies for Elizabeth, Blair, and Alleghany furnaces. and on the north side, the Alleghany Mountains contain inexhaustible beds of bituminous coal.

TIPTON is the next station. A plank road extends across the Alleghany Mountains a distance of thirteen miles, to the Clearfield Lumber Company's extensive improvements, and within four miles of the station. Coal has been found, which is said to be of an excellent quality. A very large lumber business is now doing at this station. A branch railroad, seven miles, is about to be constructed to the Alleghany coal deposits. Two miles beyond is FOSTORIA, called after William B. Foster, Jr., Vice President of the Pennsylvania Railroad Company. At this point a very large lumber business has been doing. Two miles beyond is BELL'S MILLS, a station at which a large lumber business is doing. Mr. B. F. Bell, the proprietor of the property at this point, an energetic and progressive man, is sinking an extensive well, which has been carried to the depth of 1200 feet through solid rock. Water has not yet been reached.

This is the last station till the cars stop at Altoona. At Blair's furnace, three miles below Altoona, and within half a mile of the railroad, is one of the richest banks of iron ore in the State.

ALTOONA, the great centre of the operations of the road, contains about three thousand inhabitants. Four years since its site was marked only by an old log house, which may be seen near the railroad in the western part of the town. It has been brought into existence by the Company alone. At the head of Tuckahoe Valley, and at the foot of the Alleghanies, it occupies a commanding and important position in reference to the operations of the road. East of this place the grades along the Juniata are about twenty feet to the mile. Near Altoona, ascending, they are increased to ninety-five feet to the mile. Of course the enlarged power required as well as the changes of climate, the character of the country, and many other circumstances, mark Altoona as a point where the nature of the road undergoes an essential change, and these considerations led to its being made the very heart of the Pennsylvania Railroad Company's system. The workshops of the Company are already over one thousand feet in length by seventy in width; about one thousand persons are employed. There are also extensive car-houses, blacksmith shops, carpenter shops, brass foundry, tin shop, paint shop, engine repair shops, boiler shop, store house, iron foundry, &c. &c., and two circular engine houses, one of which is six hundred feet in circumference, ninety feet high, surmounted by a dome of gigantic proportions. Here also is the house of the general superintendent, Herman J. Lombaert, and the various offices for the transaction of the business of the Company, and the LOGAN HOTEL, recently pronounced by an intelligent English traveller, to be "better than any in Europe, and equal to any in America."

This sumptuous hotel gives to the traveller unusual interest in the place. He knows that his keen appetite can be satisfied here in the most luxurious manner. This house was erected by the Company in order to insure to travellers the best accommodations at all times.

From Altoona there is a plank road, and a branch railroad to Hollidaysburg, distance seven miles, situated on the northern turnpike leading from Harrisburg to Pittsburg, at the junction of the Juniata division of the Pennsylvania Canal. To this junction, and to the creation of the county of Blair, of which this town is the seat of justice, it owes a rapid growth, having sprung in a few years from an obscure village of fifty inhabitants to a thriving town of over three thousand population. Travellers to Bedford Springs leave Altoona by the branch road, and take stages at Hollidaysburg, passing over a plank road and turnpike twenty-eight miles to that delightful watering place.

At this point we begin to climb the Alleghany Mountains, and although we ascend them at the rate of 95 feet to the mile, until we attain the height of 2,160 feet above tide water, yet we scarcely perceive any diminution of speed in the iron horse. As we ascend one mile after another, until we reach the tunnel, a distance of 12 miles from Altoona, the traveller is enchanted with the mountain scenery, and exhilarated with the idea that he is actually climbing the tall Alleghany Mountains on a railroad, at the rate of 20 to 30 miles an hour. This is one of the greatest triumphs of science and genius. The accompanying map will show the serpentine windings of the road.

After passing through the great tunnel, which is about three-quarters of a mile in length, we arrive at GALLITZIN, named after Prince Gallitzin, a Roman Catholic Priest of a noble Russian family, who settled at Loretto, in Cambria County, in 1789, and died on the 6th of May, 1840; a worthy and modest individual, who renounced the honors of his station, and devoted his time and his means to charitable purposes. It has been said of him:

"If his heart had been made of gold, he would have coined it for the poor." The tunnel through the mountain is 3,670 feet long, and the height of earth above it is 210 feet. It is handsomely arched throughout. Great difficulty was encountered in making this enormous work, but every obstacle has been happily overcome, and it now presents an object well calculated to excite the wonder of strangers, and arouse the pride of Pennsylvanians. Although passing under and through the Alleghany Mountains, at a high rate of speed, there is not the slightest danger. Indeed, accidents rarely occur where there is any appearance of peril, for the additional precautions adopted at all such places are almost absolute assurances of safety.

In descending the mountain on its western side, the traveller is cheered by the frequent view of the rapid and gradually increasing CONEMAUGH, whose limped waters rise two miles west of the tunnel, and join Stony Creek at Johnstown (a distance of 26 miles), and form the main river along which the road continues until it passes Laurel Hill and Chestnut Ridge. It may be seen sometimes on the right and sometimes on the left, hastening down the mountain as if in competition with the "iron horse." The traveller will have its cheering company for fifty miles, to Blairsville intersection. Three miles further we reach CRESSON, where Dr. Jackson has a hotel, much frequented by persons from Pittsburg during the summer, the scenery being beautiful, and the air always cool and bracing. This hotel, when completed, must become a great resort for citizens in the summer season. Its elevated situation on the Alleghany Mountain must render the air delightful in the hottest weather. The facilities of reaching the place without fatigue, the thrice a day intercourse with each of the cities, the mountain scenery, the trout fishing, and the many objects of interest in the vicinity, combine to make it a desirable retreat. The road here crosses the northern turnpike, on which, about a mile distant, and at the junction of the old Portage road, is the summit of the mountain and the town of that name. LILLYS, at the foot of Plane No. 4, on the Portage, is a water station in the woods.

PORTAGE, at the foot of Plane No. 2, is also inconsiderable.

WILLMORE adjoins the town of Jefferson, which has nearly á thousand inhabitants. A plank road connects this place with Ebensburg, the capital of Cambria County, and thence to Loretto, and northward towards Clearfield. The trade of the large section of country north of Willmore is thus brought to this point, and rapidly increasing, it promises to make this an important station.

SUMMERHILL, adjoining the old half-way house on the Portage Road, is a wood and water station.

VIADUCT. At this point the road crosses the Conemaugh at the Horseshoe Bend, by a stone viaduct of eighty feet span.

At CONEMAUGH we find an engine house, work shops, &c., of the Company, this being the western terminus of the Mountain Division. A few years since it was of but little value; but a village is rapidly growing, created by the company's works.

JOHNSTOWN was originally the point of shipment for iron brought from the Juniata to the west, and floated in flat boats down the Conemaugh, past many places where furnaces now furnish this metal at one-fourth the ancient cost. The construction of the State Canal and Portage Railroad was of great advantage to this town, but it is indebted for its present highly prosperous condition chiefly to the Pennsylvania Railroad. It lies at the junction of the Western Division of the Pennsylvania Canal and the Portage Railroad, and at the point where Stony Creek empties into the Conemaugh River. Both these streams penetrate a country rich in coal, iron, cement-rock, fine clay &c., and these mineral treasures have given to Johnstown extensive iron works and other industrial establishments. The Cambria Iron Company's establishment is one of the largest in America.

Of these works an account is to be found in "Taylor's Statistics of Coal," from which we condense a brief statement. They are situated at the confluence of the Little and Great Conemaugh Rivers, immediately below Johnstown. The property embraces about twenty-five thousand acres. The lands, with a few charcoal furnaces and other improvements, were purchased a few years ago for the sum of three hundred thousand dollars. The country has been here upheaved by some internal convulsion, and with it, on both sides of a narrow valley, the richest deposits of iron ore, bituminous coal, hydraulic cement, fire-brick clay, and limestone, in strata contiguous to each other. The principal vein, adjoining the furnace and rolling-mill (carbonate of iron), lies over the coal measures, about two hundred feet above the bed of the Conemaugh, and sixty feet above the tunnel head of the furnaces.

Besides the four blast furnaces, which have lately been enlarged and improved, and are now yielding a product of two hundred tons of pigs per week, the Company have nearly completed four other blast furnaces, for smelting with coke the iron ore taken from the face of the hill, and calculated to produce each five thousand tons of pigs per annum, making an aggregate of about twenty-eight thousand tons per annum.

They have also finished a long rolling-mill, six hundred by three hundred and fifty feet in extent, in the shape of a cross, with sixty puddling stacks, twelve heating furnaces, four scrap furnaces, &c., and which, when in full operation, will produce more than one hundred tons of railway iron per day, or thirty thousand tons per annum. This, we believe, is the largest single mill in existence. The engine for the four blast furnaces is four hundred horse power.

The Johnstown Iron Furnace, belonging to Messrs. Rhey Mathews & Co. is on a smaller scale, but its operations have been carried on during the past three or four years with extraordinary energy and signal success.

Johnstown being by far the most important point in this section of country, its people have long thought that a new county should be created out of parts of Cambria, Westmoreland and Somerset, to be called Conemaugh, with Johnstown as the county seat.

From this point there is a turnpike road to Legonier, Westmoreland County, another to Ebensberg, Cambria County, and a plank road to Somerset, population nearly five thousand.

CONEMAUGH FURNACE belongs to Rhey Mathews & Co., and has been in successful operation during the past three years. On the opposite side of the Conemaugh is Indiana Furnace, belonging to Elias Baker, of Blair Co. The population is about five hundred, and these furnaces furnish an important local trade to the railroad.

NINEVEH is at present of small account, but it is believed that it will become in time a considerable depot for the lumber and other products of the southern portion of Indiana County. There is a wood and water station at this point.

NEW FLORENCE is exclusively a railroad town, and is as yet too young to have a history. It is surrounded by a fine mineral region; there are several furnaces in the neighborhood, and it is rapidly improving. It is proposed to make it the capital of a new county to be called Legonier, and to embrace the beautiful and fertile valley of that name. From Florence roads diverge to Legonier and other points on the south, and to Indiana on the north. The Railroad Company has an extensive wood and water station at this place.

LOCKPORT, on the canal, where it crosses the Conemaugh by a very beautiful cut stone aqueduct, is the next point. John Covode, the representative in Congress from this district, an enterprising citizen of this neighborhood, has built a large and substantial brick warehouse at this place, and a considerable local trade is done chiefly from Legonier Valley. One of the late Dr. Shoenburger's iron furnaces is also located here. The roads to the south lead through Covodesville to Legonier and the southern turnpike. Just above here is a six foot vein of coal, of an excellent quality, belonging to the Westmoreland Coal Company, who are about erecting furnaces for making coke.

BOLIVAR is the seat of an extensive and most valuable fire-brick manufactory.

BLAIRSVILLE INTERSECTION. At this point the Branch Road to Blairsville (3 miles), and thence to Indiana (16 miles), diverges from the main line.

From Cresson's Station on the Alleghany Mountains to this intersection, the road has followed the Conemaugh, and the traveller cannot fail to have been alternately delighted with the beauty and awed by the grandeur of the scenery. Among the latter the most remarkable is the cut along the Pack Saddle Mountain one mile east of the intersection. The road passes through the Chestnut Ridge at a gap formed by the Conemaugh upon a narrow ledge cut out of the solid rock in the side of the mountain, at a distance of one hundred and sixty feet above the river and canal.

BLAIRSVILLE, on the right, three miles from the station, is an important town in Indiana County, on the canal and the Northern turnpike, and is the junction of the Northwestern with the Pennsylvania Railroad. This will bring Blairsville into immediate intercourse with Lake Erie and the extensive and, as yet, almost unsettled territory of Northwestern Pennsylvania, which abounds in coal, iron, and timber. The population is about three thousand, and is rapidly increasing.

INDIANA, nineteen miles from the intersection, is the capital of Indiana County, and the terminus of the Branch Railroad. It is a neat, thriving village, with excellent society, and is a pleasant place of residence. Its trade is not large, and the town has been heretofore indebted for its moderate prosperity chiefly to the public business, but it is believed that the railroad will make it a considerable shipping point, and change its peaceful tranquillity into the activity of an extensive traffic.

Returning to the main line, the next station is HILL SIDE, where the Company has an extensive depot for wood and water.

MILLWOOD is a small station, doing an inconsiderable business.

DERRY station is about one mile south of the ancient village of that name, and is an important water and wood station. Roads diverge from here to various parts of Westmoreland County.

ST. CLAIR is a new railroad town, and is improving.

LATROBE was laid out by Oliver W. Barnes, Esq., four years ago, and now contains a population of about eight hundred. It has a very extensive hotel, engine house, large warehouses, &c. There are several manufacturing establishments, including car works, iron foundry, &c. The road here crosses the Loyal Hanna by a substantial stone askew bridge. Roads diverge from Latrobe in every direction through the rich agricultural region by which it is surrounded, and a stage for Youngstown connects with every train.

BEATTY'S is a small station, connected by a turnpike two miles long, with the southern turnpike. About a mile from this place is the Benedictine Monastery and College, which numbers some two hundred students. Near this there is a young ladies' boarding school, conducted by the Sisters of Mercy. Both these institutions are located on beautiful farms, admirably cultivated, and the buildings are very extensive and complete for the purposes for which they are designed.

GEORGE'S STATION is a shipping point for a portion of the produce of the fertile country through which we are now passing.

GREENSBURG, capital of Westmoreland County, was the original seat of justice of what now constitutes most of the southwestern counties, including Alleghany; and it is a curious illustration of the rapid and decisive changes which have taken place in the west, that it is within the memory of many living men, when counsel, parties, and witnesses came from Pittsburg, during court week, to Greensburg to try their cases. Among the lawyers of these days were Ross, Breckenridge, Baldwin, Wilkins, and others, whom the traditions of the bar represent to have been of profound learning and most vigorous intellect. For many years before the construction of the railroad, Greensburg had remained stationary, but recently it has received a considerable impetus, and now feels the advantage of the rich and populous country around it. This remark applies generally to the county of Westmoreland, through which the road runs for more than sixty miles. Nature was prodigal of her gifts to this favored county—fertile soil, healthy climate, beautiful scenery, rich mines of iron, abundance of coal, gave to her people profuse means of prosperity. But want of ready and cheap access to market had rendered all these advantages comparatively valueless.

Simplicity, virtue, and contentment, are the distinguishing characteristics of the population. The railroad has opened the world to them, taught them the value of their advantages, inspired them with enterprise, and has added already enormously to the wealth of the community. It is, perhaps, not going too far to say that the additional value of the land in this county alone, is sufficient to pay the entire cost of the construction of the road.

Greensburg has always been distinguished for the refinement and intelligence of many of its inhabitants, and is a delightful summer residence, much resorted to by Pittsburgers. It contains about sixteen hundred inhabitants. Immediately east of the town is the junction of the contemplated Branch Railroad to Uniontown, Fayette County, and on the west, the Hempfield Railroad, now in process of construction, joins the Pennsylvania road, thus connecting it with the Ohio River at Wheeling.

Roads diverge from Greensburg to all parts of the country, and stages leave for Mount Pleasant, Somerset, Uniontown, Bedford, Legonier, and Cumberland, Maryland.

In the Presbyterian churchyard, on the left of the railroad as you enter Greensburg, General ARTHUR ST. CLAIR is buried. His life, obscure in its commencement, afterwards distinguished, then unfortunate, and finally full of trouble, was, nevertheless, always marked by the most scrupulous honor, and his conduct and position during the Revolution have connected his name inseparably with the history of this country. Born in Edinburgh, Scotland, he accompanied the fleet under Admiral Boscawen to America in 1755. He was a lieutenant in the British army, commanded by General Wolfe. At the close of the French war he was assigned to the charge of Fort Legonier, in the valley of that name, now the western part of Westmoreland County. One thousand acres of land having been granted to him, he laid it out in circular form, and selected almost the worst in the county—preferring, singularly enough, the barren rocks and stones of Chestnut Ridge to the fertile limestone lands of the valleys around him. He became a justice of the peace, and was then first prothonotary of the county, and at Greensburg may still be seen many records authenticated by him in the name of "our Sovereign Lord, George the Third, King of Great Britain, France, and Ireland."

> " In ancient times the sacred plough employed
> The kings and awful fathers of mankind,
> And some, with whom compared, poor insect tribes
> Are but the beings of a summer's day,
> Have held the scale of empire, ruled the storm
> Of mighty war ; then, with unwearied hand,
> Disdaining little delicacies, seized
> The plough, and greatly independent lived."

His grave was unmarked until 1832, when the Masonic Lodge of Greensburg erected a chaste and appropriate monument with the inscription: "The earthly remains of Major General Arthur St. Clair are deposited beneath this humble monument, which is erected to supply the place of a nobler one due from his country."

LUDWICK is an extensive freight station adjoining Greensburg, and being the forwarding and receiving point for that town and the surrounding country, an extensive business is done here. The Company have erected a large brick warehouse, an engine house, wood and water station, &c. A town has been laid out as an extension of Greensburg, by William A. Stokes, the able solicitor of the Pennsylvania Railroad Company, and contains already a considerable population, which is rapidly increasing.

GRAPEVILLE is also a station of the second class, and is one mile north of the village of that name on the Pittsburg turnpike.

MANOR is situated in the midst of a large body of very rich land selected by the Penns and reserved as their private property. They had many such tracts of land. Pittsburg is built on one of them. The most celebrated, perhaps, was Pennsbury, in Bucks County, which William Penn improved for his own residence, and where he hoped in peace and plenty to end his days. "Let my children," said he, "be husbandmen and housewives. This leads to consider the works of God and nature, and diverts the mind from being taken up with the vain arts and inventions of a luxurious world. Of cities and towns, of concourse beware. The world is apt to stick close to those who have lived and got wealth there. A country life and estate I like best for my children."

These proprietary manors form a curious feature in the history of Pennsylvania land titles. In 1779, the legislature vested in the Commonwealth the rights which the Penns derived from the charter of Charles II. to William Penn, declaring that the claims of the proprietaries to the whole of the soil contained within the bounds of the charter could not longer consist with the safety, liberty, and happiness of the people; that, as the safety and happiness of the people is the fundamental law of society, and it had been the practice and usage of States most celebrated for freedom and wisdom, to control and abolish all claims of power and interest inconsistent with their safety and welfare, and it being the right and duty of the representatives of the people to assume the direction and management of such interest and property as belongs to the community, or was designed for their advantage; that it had become necessary that speedy and effectual measures should be taken in the premises, on account of the great expense of the war of the revolution then going on, and the rapid progress of neighboring States in locating and settling lands heretofore uncultivated, by which multitudes of inhabitants were daily emigrating from this State. These were unanswerable reasons for abolishing rights and powers which were of a political, rather than private nature, and which were utterly inconsistent with the new position of Pennsylvania as a sovereign State. But, with strict regard for justice, this same law provided that all the private estates, lands &c., of any of the proprietaries whereof they were then possessed, or to which they were then entitled in their private several right and capacity, and all the lands known as proprietary manors, should be confirmed, ratified and established.

Thus stood the title to this MANOR and the many others selected by the Penns throughout the State; and although succeeding generations of the family have sold the most of these tracts, large and valuable bodies of land still remain which belong to them. He who seeks to purchase good land may very safely rely on the judgment of Penn and his agents, and generally feel quite safe if he can obtain a farm which is in one of the manors. The title to these lands never having been in the Commonwealth gives rise to some curious questions as to the power of the State to control them. But these discussions, now of little practical account, can hardly be considered within the legitimate range of railroad reading.

IRWIN's is a very extensive business point. In addition to the local trade of the rich surrounding country, the Coal Grove coal works, the property of Coleman & Co., of Pittsburg, are located here. The shipment of coal will exceed thirty thousand tons this year, and is rapidly increasing. It is of unsurpassed and rarely equalled quality. It is used in the gas works of Philadelphia, New York, and Brooklyn, and is considered fully equal, if not superior, to any of the foreign coals. A town is laid out here which already contains three hundred inhabitants. Roads communicate north and south from this point.

LARIMERS, 20 miles from Pittsburg, is also the centre of extensive coal operations, the works of the Westmorland Coal Co. being located here. The vein is the same as at Irvine's, and of most excellent quality. The operations of this coal company, although already approaching twenty thousand tons per annum, are still but partial and imperfect, and the quantity which is demanded for eastern supply is so large, that there is scarcely a limit to the extent of the business which will be done.

STEWART'S. At this place there is a handsome freight and passenger station.

WALL'S STATION is a depot for wood and water of the second class.

TURTLE CREEK, 12 miles from Pittsburg, is also a small station at the intersection of the Greensburg and Pittsburg turnpike.

BRINTON'S. At this point the road crosses Turtle Creek and intersects the Plank Road to Pittsburg. It is the place of departure for passengers by the Monongahela and Youghiogheny Slack Water, which extends from Pittsburg to Brownsville and West Newton. The Connelsville Railroad will connect here with the Pennsylvania Railroad. From this and the intermediate stations to Pittsburg, there are numerous trains which afford great facility to persons living near the road and engaged in business in Pittsburg. Here and at Braddock's Fields and other points in the neighborhood, coal mines and lime quarries are being opened, and an extensive business in these articles is already transacted with Pittsburg.

BRADDOCK'S FIELD is the battle ground on which Gen. Braddock was totally defeated by the French and Indians on the 9th July, 1755. At an early day in the history of this country, the French, ascending from the mouth of the Mississippi, and descending from Canada, had penetrated the west in various directions, and had made many settlements, still indicated by their names, as Vincennes, Vandalia, St. Louis, &c. Among these was FORT DU QUESNE, on the point at the junction of the Monongahela and Alleghany Rivers, on the ground now occupied by the freight depot of the Pennsylvania Railroad, appropriately named "DU QUESNE DEPOT." In June, 1775, while the war was raging which made the noblest of modern orators, WILLIAM PITT, Earl of Chatham, the greatest of English ministers, which carried British arms in triumph by sea and land around the circumference of the globe, and first taught the American colonists their growing power, an army composed of British regulars and Provincial militia marched, under command of GEN. BRADDOCK, from Cumberland to attack the French in the western wilderness. PITT, the orator, sent out this expedition. FRANKLIN, the philosopher, furnished the means of transportation. WASHINGTON, the patriot, accompanied it. Does history record any event which united in a common enterprise men such as these three?

Slowly, with difficulty, encumbered with baggage, still more encumbered by military formulas unsuited to the warfare of the woods, this army proceeded westward, and on the 9th July, crossed to the right bank of the Monongahela at a ripple about a mile below the mouth of Turtle Creek, and within ten miles of Fort du Quesne.

If the traveller will look to the left soon after passing Brinton's Station, he will see the pool made by a dam across the river. This is the spot where the army crossed. Ascending from the river, he will perceive for some distance an alluvial bottom, interspersed with ravines of various extent. These increase in number and depth as you approach and ascend the bank above the railroad. In these ravines the enemy (completely concealed by the dense forest) was posted. The British forces, "in all the pomp and circumstance of glorious war," crossed the river, marched through the level ground, and, as the advance guard approached the hills, a heavy and quick fire was opened upon them by their concealed enemy.

COL. WASHINGTON wrote to his mother from Fort Cumberland, 18th July, 1755, nine days after the battle : " When we came there, we were attacked by a party of French and Indians, whose number I am persuaded did not exceed three hundred men, while ours consisted of about one thousand three hundred well-armed troops, chiefly regular soldiers, who were struck with such a panic that they behaved with more cowardice than it is possible to conceive. The officers behaved gallantly in order to encourage their men, for which they suffered greatly, there being nearly seventy killed and wounded—a large proportion of the number we had. The Virginia troops showed a good deal of bravery, and were nearly all killed; for I believe out of three companies that were there, scarcely thirty men are left alive. Capt. Peyrouny and all his officers, down to a corporal, were killed. Capt. Polson had nearly as hard a fate, for only one of his was left. In short, the dastardly behaviour of those they call regulars, exposed all others that were inclined to do their duty, to almost certain death; and, at last, in despite of all the efforts of the officers to the contrary, they ran, as sheep pursued by dogs, and it was impossible to rally them. The general was wounded, of which he died three days after. Sir Peter Halkett was killed in the field, where died many other brave officers. I luckily escaped without a wound, though I had four bullets through my coat, and two horses shot under me. Captains Orne and Morris, two of the aides-de-camp, were wounded early in the engagement, which rendered the duty harder upon me, as I was the only person then left to distribute the general's orders; which I was scarcely able to do, as I was not half recovered from a violent illness that had confined me to my bed and a wagon for above ten days. I am still in a weak and feeble condition, which induces me to halt here two or three days, in the hope of recovering a little strength, to enable me to proceed homewards." And to his brother John he writes at the same time : "As I have heard, since my arrival at this place, a circumstantial account of my death and dying speech, I take this early opportunity of contradicting the first, and of assuring you that I have not yet composed the latter. But, by the all-powerful dispensations of Providence, I have been protected beyond all human probability or expectation; for I had four bullets through my coat, and two horses shot under me, yet escaped unhurt; although death was levelling my companions on every side of me."

It appears that Washington's estimate of the numbers of the enemy was underrated. Mr. Sparks ascertained in Paris that there were eight hundred and fifty, of whom two-thirds were Indians.

To show the difficulty and uncertainty of communication in those days, as compared with these, when thirteen hours take the traveller from Braddock's Field to Philadelphia, the following letter received by the Provincial Council on the 23d July, fourteen days after the battle, is inserted. It is curious also for the simplicity of primitive days which it discloses :—

"Sir: I thought it proper to let you know that I was in the battle where we were defeated; and we had about eleven hundred and fifty private men besides officers and others, and we were attacked the 9th day about twelve o'clock, and held till about three in the afternoon, and then we were forced to retreat, when I suppose we might bring about three hundred whole men besides a vast many wounded; most of our officers were either wounded or killed; General Braddock is wounded, but I hope not mortal, and Sir John St. Claire and many others, but I hope not mortal. All the train is cut off in a manner. Sir Peter Halkett and his son, Captain Polson, Captain Gethen, Captain Rose, Captain Tatten, killed, and many others; Capt Ord, of the train, is wounded, but I hope not mortal. We lost all our artillery entirely, and everything else.

"To Mr. John Smith and Buchannon, and give it to the next Post, and let him shew this to George Gibson, in Lancaster, and Mr. Bingham, at the sign of the ship, and you'll oblige,

Yours to command,

JOHN CAMPBELL,

Messenger.

"P S. And from that to be told at the Indian King.

"N. B. The above is directed to Mr. Smith and Buchannon, in Carlisle."

That this catastrophe was the result of the presumptuous confidence of the English officers, there can be no doubt. This stubborn Anglo-Saxon spirit, which despises danger, and sometimes makes courage rashness, is still unchanged, and has been recently shown by Lord Cardigan in the Crimea, as it was one hundred years ago by Gen. Braddock on the Monongahela.

Swissvale. A neat village has been laid out here by Mr. Swisshelm, and the beauty of the location along the entire line of the road, from Turtle Creek to Pittsburg, has attracted many persons from the latter place, who have erected residences which adorn the road and the bank of the Monongahela, and give to this region a refined and cultivated society.

Wilkinsburg, at the crossing of the Greensburg turnpike, contains about four hundred inhabitants, and is gradually improving. Here is a neat passenger station.

Homewood is called after the seat of the Hon. William Wilkins, which is about a mile distant, and is an extensive and romantic place, highly cultivated, adorned with a beautiful residence, fit for the home of the distinguished and venerable statesman, whose genius has honored his country abroad, and reflects lustre on his native State.

EAST LIBERTY is a thriving and rapidly increasing village. Many of the merchants of Pittsburg have elegant residences near this place. The station house is a fine specimen of old English architecture.

MILLVALE is near the residence of the late Hon. Harmer Denny, a distinguished citizen of Western Pennsylvania. Immediately adjoining the extensive estate of the Dennys, is the Western Pennsylvania Hospital, a noble pile of building, erected chiefly by the generosity of the people of Pittsburg on ground given by the O'Haras and Dennys.

We now enter the OUTER DEPOT of the Railroad, which consists of about twenty acres, bounded by Liberty, Morton, Ferguson, and Lumber Streets. There are seventeen parallel tracks, making an entire length of nearly six miles within the Company's grounds. Here is a locomotive and car-repair shop, altogether two hundred and forty feet long by two hundred and twenty feet. Blacksmith shop, tinner's, painter's, carpenter's, trimming shops, store houses, offices, wood houses, coal sheds, water stations, cattle yards, weigh scales and offices, local freight house, three hundred feet long by seventy-five feet wide, with offices attached ; iron and lumber yards, and a circular engine house, one of the largest in America, nine hundred feet in circumference, containing stalls for forty-four locomotive engines and tenders.

These buildings are mostly of brick, built in a substantial manner, and with considerable architectural elegance. The whole property is underdrained by brick and stone culverts. The steam engines and all the extensive machinery is of the best quality, and has all the modern improvements. An idea of the extent of the operations of a great railroad can probably be had more satisfactorily by a visit to this depot, or that at Altoona, than by any other mode. The Ninth Ward, in which these works are situated, was very sparsely inhabited before the Company located there, but it is now rapidly increasing in population, and is one of the busiest and most thriving parts of the city.

Passing through this establishment, we enter Liberty Street, and the speed of the train is diminished to about four miles per hour. The anxious precautions which are taken in passing through the streets of the city, have been so successful that no persons have ever been injured except by their own gross negligence. This remark, indeed, applies to the entire line of road, and it is a remarkable fact that *of the hundreds of thousands of persons carried in the passenger trains on the Pennsylvania Railroad, no one, sitting in his proper place, has ever been injured in the slightest degree.*

Accidents have resulted from standing on the platforms, from attempting to get on or off the cars when in motion, and from other irregularities of similar kind; but conformity to the rules of the Company, which are made for the safety and comfort of passengers, is practically equivalent to absolute assurance of security.

At Grant Street, we enter the passenger station by a curve to the left from

Liberty Street. This depot occupies a lot of ground fronting on Seventh, Liberty, and Grant Streets, five hundred feet in length by one hundred and twenty feet in width. The arrangements of the Company with the Excelsior Omnibus Line, relieves passengers from all risk and trouble as to themselves or their baggage. Before the cars reach Pittsburg, an agent passes through them, collects all baggage checks, and ascertains where the respective articles are to be delivered. He gives a check receipt for the baggage, and the passenger finds it safely delivered soon after his arrival. At the depot numerous omnibuses are in waiting on the arrival of the trains, and passengers are immediately carried to any part of the city where they may wish to go.

Opposite the passenger station the Company owns a lot of ground on the canal, Liberty, and Seventh Streets, about eight hundred feet in length by one hundred and fifty feet in width, on which a local freight depot is to be erected for the exclusive accommodation and convenience of the merchants and citizens of Pittsburg.

The railroad continues its course down Liberty Street to Marbury, when it enters the through-freight depot, a new and beautiful brick building, nearly seven hundred feet in length on Liberty Street, by one hundred and ten feet in width, and extending from Marbury Street to the Monongahela River. This is, like the outer depot, one of the sights of Pittsburg, and well worthy of a visit by persons who take an interest in the improved modes of transport and communication with which modern science and enterprise have amazed the world. Large as this building is, it will hardly be sufficient for the rapidly increasing business of the road, which, for the past year, was about 250,000 tons, and will probably, when the road is fully completed, and its advantages are thoroughly appreciated, amount to a gross tonnage of a million of tons per annum.

PITTSBURG, located in the triangle formed by the union of the Monongahela and Alleghany, which make the Ohio River, is the second city of Pennsylvania. It contains, with Alleghany and the environs, a population exceeding one hundred thousand, almost all of whom are engaged in or connected with industrial pursuits, and chiefly in the various branches of manufacturing, for which the location of the city is peculiarly favorable.

Nearly a century ago, General WASHINGTON, then a young man, standing in the forest at the junction of the Alleghany and Monongahela, predicted the fortunate destiny of what is now Pittsburg. He foresaw that the wildness of savage life would vanish at the advance of certain civilization, and that the opulence of commerce, and the refinements of intelligence would succeed the wretchedness of the original settlers, and the barbarism of the aborigines. But Washington also clearly perceived the necessity for the development of the natural resources of this favored spot by an artificial channel of communication with the east, and he was the original projector of a road to connect the waters of the Ohio with the Chesapeake Bay—the parent project of all the improvements to unite the east with the west, which have since been executed.

In 1784, Washington wrote to Governor Harrison, of Virginia, in reference to the trade of the west, and the competition even then springing up for its enjoyment: "A people who are possessed of a spirit of commerce; who see and will pursue their advantages, may achieve almost anything. In the meantime, under the uncertainty of these undertakings, they are smoothing the roads and paving the way for the trade of the western world. That New York will do the same, no person who knows the temper, genius and policy of these people can harbor the smallest doubt."

After this far-seeing exposition of our true policy, and the accurate estimate of our northern neighbors, General Washington proceeds: "Common policy, therefore, points clearly and strongly to the propriety of our enjoying all the advantages which nature and our local situation afford us; and clearly evinces that unless this spirit could be totally eradicated in other States as well as this, and every man be made to become a cultivator of the land, or a manufacturer of such articles as are prompted by necessity, such stimulus should be employed as will *force* this spirit, by showing to our countrymen the superior advantages we possess beyond others, and the importance of being on an equal footing with our neighbors." Although not directed to the particular part of the country through which we have passed, the circumstances are so similar, that every word thus written by Washington is full of instruction; and is, indeed, the wisdom which, put in action, has given Pennsylvania a great work, of which her people may well be proud.

Pittsburg is beautifully and most favorably situated. On the north flow the clear and sparkling waters of the Alleghany, which is formed by the union of several streams, some of which rise in the north of Pennsylvania, others in the southwestern part of New York. After crossing the State line, it receives successively the waters of French Creek, Clarion, Red Bank, and Kiskiminetas Rivers. The Monongahela, on the other side of Pittsburg, offers a striking contrast to the Alleghany. It is said that its name is a combination of Indian words meaning "river without an island," and such, with very unimportant exceptions, is the fact in regard to it. Its chief tributary is the Youghiogheny River. Its waters are turbid, and flow with smooth and gentle current through a fertile and beautiful country from its sources in Virginia to its mouth, a distance of one hundred and fifty miles.

The trade of these rivers is very considerable. Steamboats run up the Alleghany to Freeport, Kittanning, Franklin, and Warren. Its descending current brings down numerous and enormous rafts, which supply the larger part of all the pine timber, boards, and shingles used in the valley of the Mississippi from Pittsburg to New Orleans. Five hundred large flatboats or arks come down the Alleghany annually, loaded with corn, salt, lumber or produce, pot and pearlashes, whiskey, cheese, cabinet ware, tubs, buckets, &c., hay, oats, potatoes, hoop-poles, bark, &c. The traffic on the Monongahela is also important. Dams have been erected which make it navigable above Brownswille and Youghiogeny to West Newton. Many hundred large flatboats and steamboats run daily on these streams loaded with coal, and numerous other craft with produce, descend these rivers to Pittsburg, which in return supplies them with dry goods, groceries, iron ware, &c. Of the Ohio, nearly one thousand miles in length, "the beautiful river," as its early explorers, the French, called it, it is needless to say more than that it has the enormous commerce which belongs to one of the greatest natural highways in the world. It has but one impediment, the low water of the summer which sometimes suspends navigation for several months, creating great losses, and often involving much suffering. This is a difficulty which those competent to judge say can be obviated at moderate expense.

This river, running through part of Pennsylvania, the boundary of the great States of Virginia, Ohio, Indiana, Illinois, and Kentucky, the channel by which supplies of the most necessary importance reach besides these, the vast and numerous States bordering on the Mississippi River, or penetrated by its tributaries, is of national character in the highest sense. No one State can control it. In all that relates to it, the larger portion of the entire republic is concerned. And it is to be expected that before long the demand of the west and southwest for the efficient improvement of this their great national highway, will be so energetic and unanimous, as to secure what is so important to the interest and comfort of the many millions of people who inhabit the valley of the Mississippi.

Thus situated, it is not surprising that Pittsburg should be intimately connected with the navigation of the western rivers. The first steamboat on these waters was built here in 1811. Upwards of one hundred are now owned here—many are built every year, and among them are some of extraordinary speed and splendor. Especially is the line of packets which leave daily for Cincinnati, Louisville, and St. Louis, remarkable for punctuality, comfort, and safety, and it may well be doubted whether, considering their splendid saloons, spacious state-rooms, elegant table, excellent attendance, and motion smooth, rapid, and safe, more convenience for comfortable travelling can be found anywhere than in these packets.

More rapid communication may be found for those going west by the Pennsylvania and Ohio Railroad, which connects with the Pennsylvania Road, and by which all the important points in the west may be speedily reached. Starting from the depot, you may by this or connecting roads, reach Cleveland, Sandusky, Columbus, Cincinnati, Dayton, Indianapolis, Chicago, St. Louis, &c.

The railroad now being constructed from Pittsburg to Connellsville, thence to Cumberland, to connect with the Baltimore and Ohio Railroad— that to Steubenville, which will connect with the Indiana Road, and give a direct communication on a road of the same gauge as the Pennsylvania, through the States of Ohio, Indiana, and Illinois—that to Erie connecting

with the Northern Ohio, Michigan, and New York roads. The Alleghany
Valley Railroad, extending from Pittsburg north, and to the State line, and
joining the New York and Erie road, the Chartier's Valley road to Washington, will make Pittsburg a great artificial centre of railroad communication, as nature has made it the chief radiating point of western-waters intercourse.

The manufactories of this city are most interesting, and among these are
the cotton factories, iron foundries, steam-engine manufactories, rolling-
mills, bar and rod-iron manufactories, those of nails, glass works, steel and
brass, steam flouring-mills, steam saw-mills, rope walks, oil cloth manufactory, extensive smithshops, plough, carriage, and wagon manufactories,
shovel, spade, and fork manufactories, those for locks and small iron articles
of various sorts, establishments for boat building, and for the manufacture
of articles of leather, hats, caps, saddlery, paper furniture, and almost all
articles either for use or ornament. The coal mines on the Monongahela
side, with the railroads and other apparatus for bringing the coal to the
river, and loading it, are curious.

The public buildings of Pittsburg, of chief importance, are the Court-
house and the Penitentiary; in Alleghany, the Custom-house and Post-office,
the New Market Houses, Masonic Hall, the Catholic Cathedral, St. Peter's
Episcopal Church, the First and Third Presbyterian Churches, and the Monongahela House. The numerous bridges over the Alleghany, and the beautiful wire suspension bridge over the Monongahela, are well worthy of ex-
amination.

In this brief and imperfect sketch we have barely named some of the
objects of interest in this city. It is impossible, within the limits of this
little book to do more. We have not even named the twin sister city of
Alleghany, with her twenty-five thousand inhabitants, nor Manchester, Birmingham, Lawrenceville, &c., all contiguous to and for all practical por-
poses part of Pittsburg, and all likely soon to be consolidated into one mu-
nicipal corporation.

And so, courteous traveller, having accompanied you from PHILADELPHIA
to PITTSBURG, and pointed out what seemed likely to interest you on the
PENNSYLVANIA RAILROAD, we bid you farewell.

The following lines, written by Dr. Darwin more than half a century ago,
are now partly realized. The other part, though beautiful poetry, is a mere
flight of imagination.

> " Soon shall thy arm, *unconquered steam!* afar
> Drag the slow barge, *or drive the rapid car:*
> Or, on wide waving wings expanded bear
> The flying chariot through the fields of air.
> Fair crews triumphant leaning from above,
> Shall wave their fluttering kerchiefs as they move,
> Or warrior bands alarm the gaping crowd,
> And armies shrink beneath the shadowy cloud."

The Pennsylvania Railroad is a very important link in the chain which binds, by a direct connection, the eastern or Atlantic cities with those situated on the Ohio and Mississippi. Taking Boston and New York as points of departure for the west, say for Cincinnati, as a central point, the traveller saves in distance, in time, and in expense, one day's journey by this road. If he wish to approach any city in the west by *a direct line* from these points, he must go through the territory of Pennsylvania. He can travel, it is true, by the New York and Erie, and by the Central, but these are *circuitous routes to the west.* Their direction is to the Lakes, and lie on a circumference line, while the Pennsylvania Railroad is on a diameter line. As soon as it was ascertained that the Alleghany Mountain could be passed on a direct route to Pittsburg, without an inclined plane, the citizens of Pennsylvania, and particularly those of Philadelphia, determined that such a railroad should be made. In 1838 the first survey was made by Wm. E. Morris, an engineer. In 1841 Charles L. Schlatter was appointed by the Board of Canal Commissioners, to make a full survey for a railway from Harrisburg to Pittsburg. The first meeting of the citizens of Philadelphia, in relation to building the road, was held at the Chinese Museum, on the 10th December, 1845. It was an unusually large meeting, at which a determined spirit was manifested to prosecute this great work. Thomas P. Cope was called to the chair. A preamble and resolutions urging the importance of the work, were offered by Wm. M. Meredith, and unanimously adopted. A large committee on memorials to the Legislature, praying for an act of incorporation, and a committee of nine to prepare and publish an address to the citizens of Pennsylvania, setting forth the views and objects of the meeting, were appointed. The latter consisted of the following persons: George W. Toland, Thomas M. Petit, Henry Welsh, Isaac Hazlehurst, John M. Atwood, Robert Allen, George N. Baker, Thomas C. Rockhill, and George M. Stroud.

On the 13th of April, 1846, a law was obtained to incorporate the Pennsylvania Railroad Company. As soon as the act was passed, a large town meeting was called in Philadelphia, for the purpose of taking measures to bring the corporation into existence. At this meeting a resolution was adopted for the appointment of a committee to prepare an address to the citizens, urging the measure. On this committee the following named gentlemen were appointed by the chairman, Thomas P. Cope, viz: Job R. Tyson, David S. Brown, John Grigg, Thomas Sparks, George N. Baker, Richard D. Wood and James Magee. This committee prepared an address which was issued in pamphlet form, and was extensively published in the newspapers. It was understood to be from the pen of its chairman, Job R. Tyson. It met with a warm response from the citizens. Private and corporate subscriptions were soon obtained, particularly one from the city, of two and a half millions of dollars. This subscription gave an impulse to the enterprise that left no longer any doubt about its success. The first Board of Directors consisted of the following gentlemen, most of whom had been active in promoting this great work, viz: S. V. Merrick, Thomas P. Cope, Robert Toland, David S. Brown, James Magee, Richard D. Wood, Stephen Colwell, George W. Carpenter, Christian E. Spangler, Thomas T. Lea, William C. Patterson, John A. Wright, and Henry C. Corbit. First officers—S. V. Merrick, President; Oliver Fuller, Secretary; George V. Bacon, Treasurer; J. Edgar Thomson, Chief Engineer; William B. Foster, Jun., Associate Engineer, of the Eastern Division; Edward Miller, of the Western.

Although the cost of the Pennsylvania Railroad is great, the capacity of it is great also, and capable of producing a large remunerating net income. It is calculated that the tonnage of the road, when completed with double track, can be increased to a million of tons per annum, independent of the passenger business. This amount of tonnage, with the passenger business, will require 300 engines of 200 horse-power each. Allowing two-thirds of the engines to be in daily use, the quantity of water consumed by them per day would be 1,200,000 gallons, which is nearly one-third of the amount consumed daily by the inhabitants of the city of Philadelphia. The conversion of this enormous amount of water into steam would require 810 tons of coal, or 2,000 cords of wood per day. The chopping, hauling, and preparing this amount of wood would give employment to 3,000 laborers. The number of eight-wheeled cars which would be required to accommodate one million of tons per annum, carried over the whole road at a speed of ten miles per hour, would be 4,500. The total number of employees of every description required, would not be less than 4,050. The income of the road, thus equipped, and transacting such a business, would be, at low rates, $5,000,000. The population supported by such a business directly and incidentally, may be estimated at 50,000 persons, not including those supported by dividends on capital, or in the preparation of iron, lumber, and other materials, which would swell the aggregate to perhaps 100,000 persons.

☞ Any person who may detect an error either in the book or map, or who may desire to have any alteration or addition, or who may be possessed of any interesting information respecting any place through which the road passes, would confer a favor by communicating the same to the Secretary of the Company, Edmund Smith, before the issuing of a second edition.

Printed by BoD™in Norderstedt, Germany